WHAT WOULD HAPPEN IF...

A SUPERVOLCANO ERUPTED?

Written by Izzi Howell

Illustrated by Paula Bossio

WORLD BOOK

www.worldbook.com

READING TIPS

This book asks readers to ponder the question *what would happen if a supervolcano erupted?* Readers will discover the threats volcanoes pose for our world and how a supervolcano eruption would affect life across our planet. Use these tips to help readers consider the ripple effects of certain actions and events.

Before Reading

Explain to readers that this book uses cause and effect to show how a change in one part of the world can affect the environments and living things throughout the rest of the world. Cause and effect can help us think about why things happen the way they do. It can also help us think about what might happen in the future. Encourage readers to be on the lookout for examples of a cause and effect structure as they explore what would happen if a supervolcano erupted.

During Reading

Discuss with readers how some actions and events have multiple causes and others have multiple effects. Explain that it can be tricky to keep all the if/then scenarios straight in our minds, so it can be helpful to create a visual guide. Encourage readers to draw and add notes to their own cause and effect maps like those found on pages 8, 10-11, and 30-31.

After Reading

After finishing the book, discuss with readers how their understandings and opinions of volcanic eruptions and their widespread effects have changed. Additionally, you can have readers respond to the comprehension questions included on page 46 and can complete the Chain of Events activity on page 47 to further extend the learning.

Visit **www.worldbook.com/resources** for additional, free educational materials.

There is a glossary of terms on pages 44–45. Terms defined in the glossary are in boldface type that **looks like this** on their first appearance on any spread (two facing pages).

Contents

Supervolcano danger 4
Volcanoes 101 6
Enormous eruptions 14
A supersized disaster 20
After the eruption 26
Supervolcano safety 32
Conclusion 40
Summary 42
Glossary 44
Review and reflect 46

Supervolcano danger

You've probably heard of a volcano, but how about a supervolcano? Sounds a little scary, right? Supervolcanoes are capable of gigantic, incredibly destructive eruptions that are far more powerful than those from a normal volcano. Should we be worried?

In short, not really! Supervolcanic eruptions are incredibly rare. This is because there are very few volcanoes on Earth that are capable of such enormous explosions. The last one happened around 26,500 years ago, when huge amounts of **lava,** gas, and **ash** exploded out of the Taupō volcano in New Zealand.

DID YOU KNOW? Of the 1,500 **active** volcanoes on Earth at the moment, only around 20 are believed to be supervolcanoes.

It's not me, promise!

Lake Taupō, the largest lake in New Zealand, formed in the huge **caldera** (crater) left behind when the supervolcano erupted and collapsed in on itself.

Even though it's very unlikely that a supervolcano will erupt soon, it's still important for us to be prepared. Scientists are constantly gathering data from volcanoes to calculate when they might erupt next and how big the eruption might be. This information helps us **monitor** supervolcanoes and could warn us of a potential supervolcanic eruption. We can also study clues left behind by previous supervolcanic eruptions to figure out what might happen if a supervolcano erupted again.

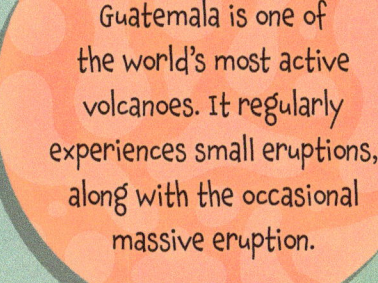

Volcán de Fuego in Guatemala is one of the world's most active volcanoes. It regularly experiences small eruptions, along with the occasional massive eruption.

THINK ABOUT IT!

Even though a supervolcanic eruption probably won't happen anytime soon, people are still very interested in supervolcanoes! Why do you think that is?

Supervolcanic eruptions are very rare, but they have happened before and will eventually happen again. When, where, and how? No one really knows! If a supervolcano did erupt soon, what would happen? Are we ready for it? Read on to find out!

Volcanoes 101

If you want to understand supervolcanoes, first you need to come to grips with volcanoes! A volcano is a place where molten rock **(lava/magma), ash,** and gas from deep underground burst from Earth's surface. Over time, a mountain of rock and ash can build up on this spot. This mountain is also known as a volcano.

FUN FACT! Molten rock is called magma when it is underground and lava when it is above ground following an eruption.

Hey magma!

Hey lava!

Mount Fuji in Japan is technically an active volcano! It hasn't erupted since 1707.

Volcanoes can be described as **active, dormant,** or **extinct.** An active volcano has erupted recently or is likely to erupt again soon. A dormant volcano hasn't erupted for a long time but may erupt at some point in the future. An extinct volcano will almost definitely never erupt again.

DID YOU KNOW?

At 13,677 feet (4,169 meters) tall, Mauna Loa in Hawaii is the world's largest volcano.

Mount Stromboli, off the coast of Italy, has been erupting continuously for around 2,700 years!

Over 80% of the rock on Earth's surface comes from volcanoes.

The eruption of Krakatau in Indonesia in 1883 was heard 3,000 miles (4,800 kilometers) away. It was the loudest sound ever recorded.

There are volcanoes in space! Mars is home to the largest volcano in the solar system, Olympus Mons, which measures 370 miles (600 kilometers) in diameter.

THINK ABOUT IT!

Where is the nearest volcano to your hometown? If you don't live near a volcano, what do you think it would be like to live close to one?

Volcanic eruptions can be very dangerous. Hot lava and ash can burn and bury people and buildings. Clouds of gas and ash released during the eruption can cause problems in the surrounding area and even farther away. However, volcanoes also have some advantages. In volcanic areas, ash and lava break down to **fertilize** the soil, making it easier to grow **crops**. The heat from magma can also be used to generate **geothermal** energy.

VOLCANOES 101

Volcanoes can erupt in different ways. Some eruptions are calm, with slow streams of **lava**. Other volcanoes have powerful eruptions and hurl out huge amounts of gas, **ash,** and rock.

However, all volcanic eruptions start off in the same way.

BEFORE THE ERUPTION

Magma builds up underground in an area called a **magma chamber.**

⬇

When the pressure becomes too great, magma is forced up through the volcano and out onto the surface.

⬇

Pressure increases as more magma fills up the magma chamber.

⬇

The volcano explodes!

DURING THE ERUPTION

Huge clouds of ash and poisonous gas are released.

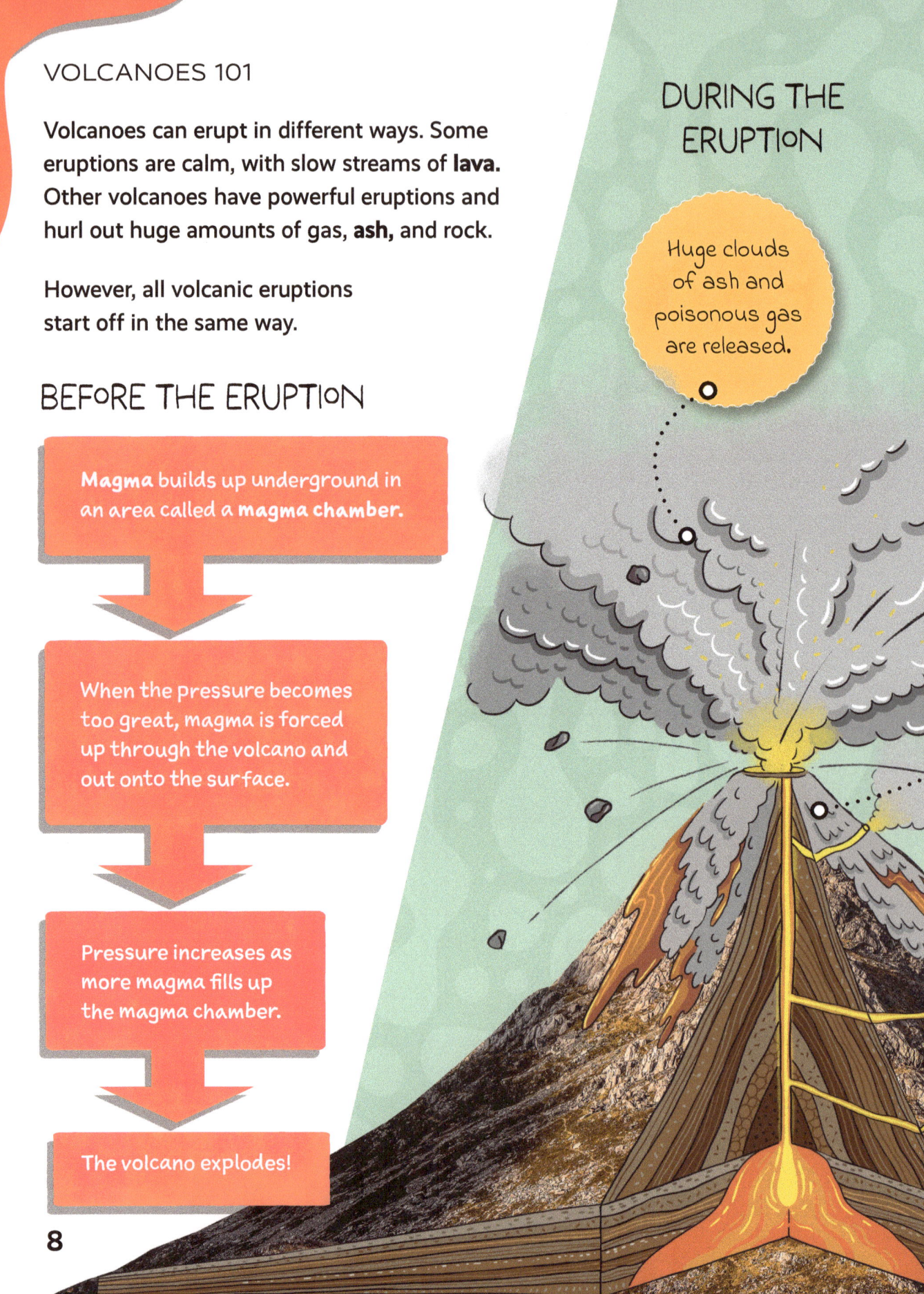

Extremely hot **pyroclastic flows** made up of ash, gas, and solidified lava rush out of the volcano at speeds of up to 650 feet (200 meters) per second.

Chunks of rock are thrown out of the volcano when gas explodes. These usually fall near the vent, so they aren't particularly dangerous (unless you are near the vent, in which case you'd have bigger problems to worry about!).

DID YOU KNOW?

Pyroclastic flows can reach temperatures of up to 1100 °F (600 °C)!

Some magma can escape through smaller side vents.

This runny lava flows gently out of the volcano. Lava usually moves at less than 6 miles (10 kilometers) per hour, so most people would be able to escape it easily.

THINK ABOUT IT

Which parts of a volcanic eruption do you think are the most dangerous? Use the information on this page, then turn to the next to check your answer!

VOLCANOES 101

Volcanic eruptions can have a big impact on the area around the volcano and sometimes even further afield. Let's take a look!

Runny lava flows

Pyroclastic flows

When **lava** flows cool down, they harden into solid rock. Anything under this rock, such as buildings, roads, and other structures, will be trapped underneath. It can be a long, slow, and expensive process to remove the hardened lava. Eventually, plants can grow back through some areas of lava, and the original habitat returns.

Pyroclastic flows have the double danger of being incredibly hot and incredibly fast. They bury and burn anything in their path and can even cross rivers. Many deaths connected to volcanic eruptions are caused by pyroclastic flows.

Lahars

Dangerous mudflows called **lahars** can form if loose **ash** mixes with rain or melted snow. These pick up anything in their path, including trees and large rocks. When the mud dries, it becomes as hard as cement.

Clouds of ash and poisonous gas

Gas

Poisonous gas can hurt and even kill nearby people and animals. They can also poison nearby rivers and lakes.

Ash

Ash can do great damage to airplane engines if it gets inside. Even small amounts of ash are very dangerous to airplanes, so airplanes can't fly anywhere near the site of certain eruptions.

If enough sulfur dioxide is released into the **atmosphere,** it can block out heat from the sun. This can make it much colder on Earth (see pages 26–27).

People and cargo can't travel by airplane, which limits trade and the movement of people (see pages 28–29).

Crops can fail if temperatures are much lower than usual, which affects our food supply.

Tsunamis

If large amounts of volcanic material fall into the sea following an eruption, it can trigger a massive wave known as a **tsunami.**

Caldera

If enough material is removed from the volcano, it can collapse. This creates a huge crater known as a **caldera.**

VOLCANOES 101

Volcanoes form above areas where **magma** is produced in the **mantle,** the rocky layer beneath Earth's **crust.** This happens all over the world, both on land and underwater on the seabed. Most volcanoes are located along the edges of **tectonic plates** (huge pieces of Earth's crust). Some also form above hotspots in the middle of tectonic plates.

Earth's crust is made up of around 30 tectonic plates, which are always slowly moving around. When two plates move away from each other or crash into each other, it's volcano time!

DID YOU KNOW? The spreading of tectonic plates on the seabed creates about 1 square mile (2.4 square kilometers) of ocean crust a year.

When two tectonic plates move away from each other, magma rises through the gap created and flows onto the surface. This mostly happens on the seabed. The **lava** hardens to create new rock on the edges of the two plates. Over time, this makes the seabed larger.

Don't touch the red stuff, trust me!

When two tectonic plates bump into each other, one plate is usually forced under the other. The plate that is pushed underground melts in the mantle, creating extra magma. This extra magma can travel up through the plates and burst out of the surface as a volcano.

THINK ABOUT IT!
Earthquakes are common along the boundaries where two tectonic plates move toward each other. Why do you think that is?

Bump!

Eurasian Plate
Philippine Plate
North American Plate
Pacific Plate
South American Plate
Nazca Plate
Indian-Australian Plate
★ Volcano
Antarctic Plate

Tectonic plates meet each other all around the edges of the Pacific Ocean. In this region, known as the Ring of Fire, volcanoes and earthquakes are very common.

FUN FACT!
Seventy-five percent of volcanoes are found in the Ring of Fire.

Volcanoes can also form in the middle of tectonic plates in areas called hotspots. Here, unusually hot rock rises through the mantle and can erupt out of the crust.

The Hawaiian Islands were created by volcanic activity above a hotspot.

Enormous eruptions

So what's the difference between a volcano and a supervolcano? It all comes down to something called the Volcanic Explosivity Index (VEI). This scale measures the size of a volcano's eruption, considering how much rock, **ash,** and gas come out during the eruption, how far the volcanic material travels, and how long the eruption lasts.

The VEI goes from 0 (the weakest eruption) to 8 (the strongest eruption). Each level on the scale is 10 times more powerful than the previous level. If a volcano experiences a level 8 eruption, it's a supervolcano! Let's see how some famous eruptions are ranked on the VEI.

LEVEL 4 ERUPTION

The 2010 eruption of Eyjafjallajökull in Iceland was small but mighty! It created huge ash clouds that reached almost 7 miles (11 kilometers) up into the **atmosphere.** The ash spread across Europe, disrupting air travel there for a week (see page 11).

LEVEL 5 ERUPTION

The eruption of Mount Vesuvius in Italy destroyed the Roman cities of Pompeii and Herculaneum in A.D. 79. Thousands of people died when **pyroclastic flows** covered the area around the volcano. The Roman cities were buried under ash, preserving many ancient buildings and objects.

LEVEL 6 ERUPTION

The 1991 eruption of Mount Pinatubo in the Philippines was one of the largest volcanic eruptions of the 20th century. Rock and ash were shot 22 miles (35 kilometers) into the sky. However, thanks to careful **monitoring** of the volcano, people living nearby were **evacuated** in time.

LEVEL 7 ERUPTION

In 1815, the eruption of Indonesian volcano Mount Tambora became the most powerful volcanic eruption in recorded history. It ejected a staggering 36 cubic miles (150 cubic kilometers) of ash and rock, which buried the surrounding area. Sadly, tens of thousands of people died.

The large amounts of ash released by Mount Tambora resulted in "a year without summer," in which parts of the Northern Hemisphere experienced summer snow and frost in the year following the eruption.

THINK ABOUT IT!

If a level 7 eruption can do that much damage, what do you think a supervolcanic level 8 eruption would be like? Come up with your own ideas, then check them on pages 16–19.

It's a level 5, for sure!

Hi! We're Stephen Self and Christopher G. Newhall. We're the **volcanologists** who created the VEI scale in 1982. Before that, there wasn't any way to compare eruptions. We love to research big eruptions that rank high on the chart. If we can predict which volcanoes will have massive eruptions, we can make sure we're prepared!

ENORMOUS ERUPTIONS

Since there hasn't been a supervolcanic eruption in tens of thousands of years, we don't have any accounts of what they are like. However, supervolcanoes from the past have left plenty of clues that we can use to find out what happens during these catastrophic events.

Have you ever heard of the Yellowstone supervolcano? This large volcanic area in the Western United States has experienced not one, but two level 8 eruptions (2.1 million years ago and 640,000 years ago) and one level 7 eruption (1.3 million years ago). These events greatly affected the surrounding area and all of North America.

Today, the beautiful Yellowstone National Park sits on top of the huge **calderas** left behind by Yellowstone's three gigantic eruptions. The size of these calderas gives us a sense of just how massive these eruptions were. The caldera created by its most recent eruption measures an incredible 47 miles (76 kilometers) long and 28 miles (45 kilometers) wide.

Yellowstone is still an **active** volcanic area. The heat from its **magma chambers** creates amazing hot springs and geysers, like this one here.

16

This is outrageous!

Yellowstone's most recent eruption also released a huge amount of ash. The surrounding area was buried in up to 660 feet (200 meters) of ash. The wind also carried **ash,** dropping several yards of ash over much of western North America. **Geologists** have even found layers of ash from Yellowstone in Louisiana, which is nearly 1,500 miles (2,400 kilometers) away.

There has been a lot of discussion over the years about whether Yellowstone will be the next big supervolcano to erupt. Yellowstone's two magma chambers still contain a lot of **magma,** and it has experienced several major eruptions before, so why not again?! There's no need to worry, though. A future Yellowstone eruption is possible, but it is very unlikely to happen in the next few thousand years.

DID YOU KNOW?
Scientists believe it's much more likely that Yellowstone will experience a low-level eruption with runny **lava** flows than a gigantic supervolcanic eruption.

Yellowstone will probably erupt again in the distant future. This is an artist's impression of what it will look like.

FUN FACT!
One of Yellowstone's magma chambers contains enough magma to fill up the Grand Canyon 11 times!

Looking at the site of Mount Toba in Sumatra, Indonesia, today, it's hard to imagine that the largest volcanic eruption in the past 2 million years once took place here. Around 74,000 years ago, Mount Toba exploded and collapsed in a catastrophic eruption, which released a staggering 670 cubic miles (2,800 cubic kilometers) of **ash** and rock. To put that into context, its eruption was about 50 times larger than the massive level 7 eruption of Mount Tambora in 1815!

This diagram shows the area covered by volcanic ash following the eruption of Mount Toba.

Mount Toba covered areas as far away as India in about 20 feet (6 meters) of ash.

1,860 miles (3,000 km)

ENORMOUS ERUPTIONS

Following its eruption, Mount Toba collapsed, creating a 20 by 60-mile (30 by 100-km) **caldera**. This filled with water to become Lake Toba, which is the largest volcanic lake in the world.

FUN FACT!
The amount of **magma** released by Mount Toba was about twice the volume of Mount Everest!

As well as destroying most of the landscape and living things across Sumatra and much of Indonesia, ash and gas released by Mount Toba also led to a severe **volcanic winter** (see pages 26–27). Evidence found buried deep in polar ice suggests that worldwide temperatures fell by 5.4 to 9.0 °F (3 to 5 °C) and as much as 18 °F (10 °C) in some places.

Bad news—scientists believe that there is still a large amount of magma under the Toba caldera. However, the good news is that its next eruption probably won't happen for thousands of years. Whew!

When Earth's next supervolcanic eruption finally happens, scientists predict that it will be as explosive, destructive, and powerful as these previous eruptions. Turn the page to see what that might look like for our planet.

THINK ABOUT IT!
Humans lived on Earth at the time of the Mount Toba eruption. How do you think they felt when they noticed the effects of the eruption?

19

A supersized disaster

A supervolcanic eruption will almost definitely not take place during our lifetimes, but a supervolcano will erupt again at some point in the future.

We have no way of knowing for sure which volcano will experience the next supervolcanic eruption, but we do have a list of likely candidates. Scientists believe that it's possible that a previous supervolcano, such as Yellowstone or Mount Toba, could experience another supervolcanic eruption again.

It's hard to believe that this green, rocky landscape was once the site of one of the most powerful events on Earth. The supervolcanic eruption that created La Garita **Caldera**, United States, was 10,000 times more violent than that of Mount St. Helens. Luckily, this area is no longer volcanic, and so won't erupt again. Whew!

The next supervolcanic eruption could also come from a volcano that has only ever had lower-level eruptions. Scientists have identified a group of volcanoes around the Aira Caldera in Japan and the Campi Flegrei volcanic area in Italy as places where a first-time supervolcanic eruption could be possible.

The Sakurajima volcano is located inside the Aira Caldera in Japan. It has been erupting almost constantly since 1955.

Almost eruption time!

DID YOU KNOW?
Around 1.5 million people live in and around Campi Flegrei in Italy. If the supervolcano sprang into action, they'd all need to be **evacuated!**

Hi! I'm Colin Wilson, a British **volcanologist.** During my research, I've realized that supervolcanoes can erupt in so many different ways. Their eruptions look different, start for different reasons, and last for different amounts of time. For this reason, it's very hard to predict how supervolcanoes will erupt in the future and which one will erupt next. However, we do know lots about predicting when eruptions will happen, so we'll be able to warn people in good time!

THINK ABOUT IT!
Would you rather know about a supervolcanic eruption years in advance, giving you lots of time to worry, or learn about it closer to the time and have less time to prepare?

Let's take a look at one supervolcano—Yellowstone—and see what we could expect following another supervolcanic eruption.

Lava from the volcano wouldn't travel too far. It would probably stay within the national park. However, the scorching hot rock would destroy everything in its path, including roads, plants, and park facilities. It would also kill many animals and could trigger wildfires.

Pyroclastic flows from Yellowstone would reach much farther. Anyone living in this area would need to be urgently **evacuated** due to the speed of the flows.

Scientists believe that **ash** would probably be the greatest threat following an eruption of Yellowstone. Areas within a 150-mile (240-kilometer) radius would be covered in more than 40 inches (1 meter) of ash. The weight of this ash would destroy all buildings and kill plants and animals. This massive area would need to be evacuated.

DISTANCE FROM VOLCANO

But that's not all! Ash from Yellowstone would be carried across North America by the wind. Most of the United States would experience some ashfall. Even a few inches of ash can destroy farms, block **sewers,** break power lines, and cause breathing problems.

Ash thickness (high to low)

These **crops** will not survive after being buried in ash.

DID YOU KNOW?

Scientists believe that a supervolcanic eruption from Yellowstone would create a massive umbrella ash cloud, which would spread out in all directions.

THINK ABOUT IT!

How many people do you think would need to be evacuated if Yellowstone erupted? Why would this be difficult?

A SUPERSIZED DISASTER

If the next supervolcano to erupt is near the coast, the risk level will increase even further! When a volcano erupts into water, it often creates an enormous and deadly **tsunami.** Based on the tsunamis that have followed lower-level eruptions, such as Krakatau, scientists believe that a supersized eruption will set off a supersized wave!

Tsunamis can be triggered when a coastal volcano collapses or releases large **pyroclastic flows,** or when an underwater volcano erupts. They travel out in circles from the site of the eruption at great speed. In deep water, they can move as fast as 600 miles (970 kilometers) per hour.

DID YOU KNOW?
In deep water, a tsunami can move as fast as a jet aircraft!

It's very hard to spot tsunamis out at sea. Even boats rarely notice sailing over them. But as soon as they approach the shore, their truly terrifying size is revealed. Some tsunamis have been known to reach 100 feet (30 meters) above sea level and travel up to two-thirds of a mile (1 kilometer) inland.

Large, powerful tsunamis can cross oceans. This means that we'd not only need to **evacuate** people in the path of the eruption, but also evacuate people from many coastlines. If these areas weren't evacuated, many people would lose their lives. Even if everyone was successfully evacuated, massive areas of coastline would be destroyed by the waves.

DID YOU KNOW? The eruption of the underwater volcano Hunga Tonga-Hunga Ha'apai near the Pacific island of Tonga in 2022 created tsunami waves that traveled as far as New Zealand, Peru, Japan, and California, U.S.

Geologists monitor the eruption of Hunga Tonga-Hunga Ha'apai. As well as triggering tsunami waves, nearby Tongan islands were also affected by falling **ash**, which damaged **crops** and poisoned their drinking water.

25

After the eruption

It's tempting to think that we could all relax after a supervolcanic eruption, but sadly, that's not the case! Even normal volcanic eruptions can have major long-term consequences on our planet. A supersized eruption would have an even bigger impact. Its effects would probably last for years.

Do you remember how some volcanic eruptions changed Earth's climate? This is because volcanoes release large amounts of **ash** and sulfur dioxide into the **atmosphere.** The sulfuric acid mixes with water to create tiny **particles** that are rich in sulfur. These particles reflect light and heat from the sun back out into space.

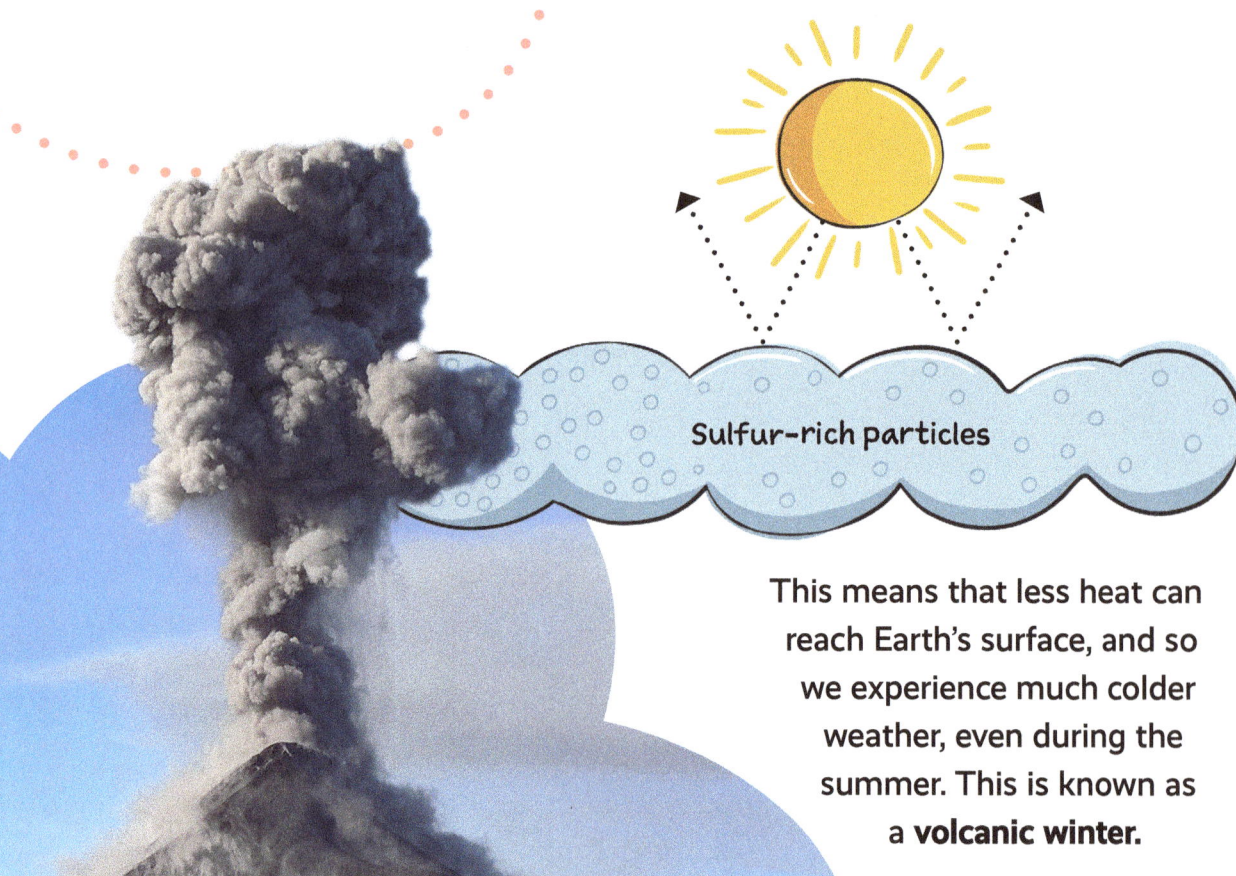

Sulfur-rich particles

This means that less heat can reach Earth's surface, and so we experience much colder weather, even during the summer. This is known as a **volcanic winter.**

Following the level 6 eruption of Mount Pinatubo in 1991, global temperatures dropped for two years. Considering how much a lower-level eruption can affect Earth's climate, it's safe to assume that a future supervolcanic eruption would also be followed by years of much colder temperatures.

During this time, many **crops** wouldn't be able to grow properly without warm weather. This would lead to a massive decrease in our food supply. In some places around the world, people would probably go hungry, and some might die.

Delicate plants that produce fruit in the summer, such as tomatoes, would be very hard to grow during a volcanic winter.

Wild plants would also struggle to grow, which would result in food shortages and hunger for wild animals. We'd probably see decreases in the populations of plant-eating animals, followed by a drop in the number of animals that eat plant-eating animals. In the end, entire **ecosystems** would be affected.

THINK ABOUT IT!

What would a year without summer be like for you? What would you miss?

AFTER THE ERUPTION

If the level 4 eruption of Eyjafjallajökull in 2010 could make air travel across Europe grind to a halt for a week, just imagine the disruption that a level 8 supervolcanic eruption could cause! It's likely that **ash** would remain above the volcano and even further afield for weeks, months, or even a year after the eruption. So, what would that mean for air travel?

When we think of traveling by airplane, many people think of vacations or trips to faraway places. But airplanes are used for so much more. Some people work for international companies and must travel around as part of their job. Others fly back and forth regularly to help and support friends and family members. Some of these trips could be made by car, train, or boat, but they would take longer.

Our grocery stores are filled with fruit, vegetables, seafood, and meat that were grown and produced thousands of miles away. Some of these foods wouldn't survive a long trip by boat and would go bad before they reached their destination. Without airplanes, we'd miss out on many foods that we currently eat every day.

Out-of-season strawberries and other fruits and vegetables are flown around the world, so people can enjoy them all year round.

DID YOU KNOW? Transporting food by airplane creates around 50 times more carbon emissions than transporting it by boat.

Thanks to the internet, it's so easy to buy things from other countries. Thanks to airplanes, we can quickly receive these purchases of clothes, electronics, books, toys, and so much more. Previously, it took weeks or even months for international mail to be transported by boat. If airplanes couldn't fly, we'd be back to the slow days of snail mail!

As well as being inconvenient, the loss of international travel would ruin many businesses. Farmers wouldn't be able to sell their **crops,** stores wouldn't be able to sell their products, and supermarkets wouldn't have any food to sell. Many people would struggle to make enough money to survive.

Over 30 million flights take place every year. Imagine if none of those flights could take off!

THINK ABOUT IT!

Have you ever traveled by plane? Would you be able to make the same trip by another mode of transportation? How would traveling in this way be different from traveling by plane?

AFTER THE ERUPTION

What would would be the longer-term effects of a supervolcanic eruption?

After **lava** and **pyroclastic flows** stopped erupting from a supervolcano, clouds of **ash,** sulfur dioxide, and other gases would linger in the air for a long time. What would this mean for our planet?

Airplanes can't fly through any areas with large quantities of ash in the sky. If a supervolcanic eruption was large enough, it might release enough ash that all planes in a certain area, or even everywhere on Earth, would have to be grounded for weeks, months, or even years.

Certain foods, products, and other cargo couldn't be transported by air. We'd have to eat more locally produced foods and wait longer for products sent from other countries.

People wouldn't be able to travel by airplane. They'd have to travel by car, boat, or train instead. Journeys would take much longer and might become much more expensive.

Many businesses that depend on air travel would suffer. If they couldn't send their products by air—or travel by air to do business—they might have to close down. As a result, some people would struggle to make enough money or even lose their jobs.

Tiny sulfur **particles** would gather in the **atmosphere.** They would reflect heat from the sun back into space.

THINK ABOUT IT!

Which of these consequences would be the most difficult for you personally, and why? Which do you think would be the biggest problem for the whole world?

Again?

Less heat would reach Earth, which would cool down our planet and lead to a **volcanic winter.** Temperatures would probably be colder than normal for years after the eruption.

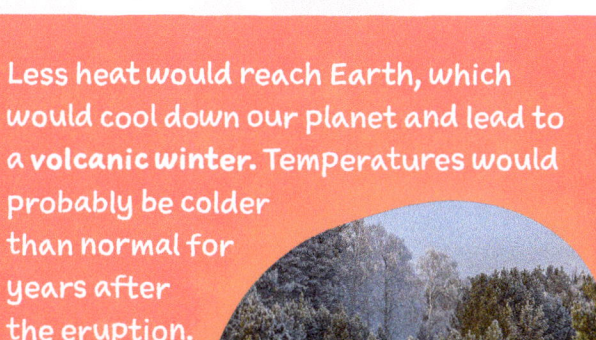

Many wild plants wouldn't grow well in the cold weather. Many would die.

Many **crops** wouldn't grow well in the cooler temperatures. Some might not grow at all, leaving us with much less food to go around.

If wild plants didn't grow, there wouldn't be enough food for plant-eating wild animals. Many would starve, and their population would go down.

I'm so hungry!

Supervolcano safety

Even though the chances of a supervolcanic eruption are extremely low, it's still better to be prepared! Thanks to the hard work of **volcanologists** and other scientists, we will probably be able to predict if a supervolcano is about to explode and get everyone to safety. Volcanoes can erupt in many ways, so it's important to learn as much as we can about all the possible outcomes.

In the 1990's, a team of scientists put together a list of 16 volcanoes that had previously experienced huge, destructive eruptions and that were close to populated areas. This list included Mount Etna in Italy, Mount Rainier in the United States, Unzen in Japan, and Merapi in Indonesia. They thought that focusing research on these specific volcanoes would help protect more people from future eruptions.

Mount Rainier in the U.S. is a potentially highly dangerous volcano because it is so close to the city of Seattle, and scientists believe it will erupt again relatively soon.

DID YOU KNOW?

In 1792, Mount Unzen erupted and partially collapsed, creating a massive landslide and **tsunami**.

Over the next 10 years, volcanologists researched how these volcanoes erupt and planned what to do in the event of an eruption. Some of these volcanoes had barely been studied before, so they learned lots of new information. They also created individual plans for each volcano for what to do in the event of an eruption.

Their work was quickly put to the test in 1992 when **lava** from Mount Etna started heading toward the nearby town of Zafferana. Scientists knew that the lava came from one underground tube inside the volcano, so they blocked the tube with huge concrete blocks to stop the flow! This strategy worked, and the village was saved—hooray!

Town of Zafferana

Hardened lava from Mount Etna

You can see just how close the town of Zafferana was to being destroyed by lava from Mount Etna.

Taking a similar approach with supervolcanoes will help us prepare for a future eruption. If we identify and study the volcanoes that are most likely to experience a level 8 eruption and put plans in place *now*, the chances of more people surviving a future eruption will be much higher!

THINK ABOUT IT!

Do you think a similar approach would work with other natural disasters, such as earthquakes, wildfires, or storms? Why or why not?

SUPERVOLCANO SAFETY

There's no way of knowing for sure exactly when a volcano will erupt, but there are many clues that scientists look for. **Monitoring** volcanoes in this way won't stop them from erupting, but it will help keep people safe.

Just like our stomachs bulge after eating a big meal, volcanoes can also bulge when they are full to the brim with **magma!** The extra pressure from magma building up in the **magma chamber** can make volcanoes change shape in the run-up to an eruption.

Gas meter on a plane

Earthquake sensor

Earthquakes are also a massive clue that an eruption is on its way! This is because the movement of magma can put pressure on the rock and cause it to move or break, creating an earthquake. **Volcanologists** use seismographs to track how many earthquakes take place near a volcano and how strong they are. They look for patterns in the data to help them predict eruptions.

Scientists monitor the gas released by volcanoes, since the amount and types of gas released can change just before an eruption. They place gas-measuring instruments on airplanes and fly them just above the volcano or place the instruments in a safe place near the volcano.

Gas meter on the ground

FUN FACT!
Hot springs near a volcano often get hotter right before an eruption!

Hi! I'm Ji Gao, a Chinese scientist who studies volcanoes and rock. At the moment, scientists can predict eruptions far in the future, or a few days or hours in advance. Ideally, we'd like to be able to confidently predict eruptions a few weeks or months in advance, so that we have time to **evacuate** and prepare. I've recently worked on a project where we used a special camera to measure magma levels underneath the Weishan volcano in China. This is another tool that volcanologists can use to make their predictions more accurate and save more lives!

35

SUPERVOLCANO SAFETY

Evacuation is the best way of keeping people safe in the event of a volcanic eruption. Scientists can usually predict when an eruption will happen, giving people enough time to move out of the danger zone.

However, evacuation isn't always as simple as that. It can be hard to convince people to go, especially if they have had false alarms before. Some people might not want to leave their homes and possessions behind. Others might want to evacuate but can't afford to travel or stay elsewhere. These are all important issues that governments need to consider when putting together volcano safety plans.

Plymouth, the former capital of the Caribbean island territory of Montserrat, was totally destroyed by **pyroclastic flows** from the Soufrière Hills volcano in the 1990's. Many lives were saved by evacuating Plymouth, but the town itself could not be saved. Today, it is a **ghost town**.

DID YOU KNOW?

Before Soufrière Hills started erupting, about 13,000 people lived on Montserrat. Many did not return after being evacuated to nearby islands. Only around 5,000 people live there today.

36

Hi! I'm Jenni Barclay, a British **volcanologist**. I'm passionate about finding ways to reduce the risk of volcanoes. When teaching people about volcano safety, science isn't enough. It's also important to understand people's culture and beliefs and include them in your work. I run lots of projects in South America and the Caribbean that help to share information about volcanoes with local communities. I also set up **citizen science** projects there, so people can **monitor** volcanoes themselves!

A powerful supervolcanic eruption would affect such a large area that a huge number of people would need to be evacuated. For example, if Yellowstone erupted, many U.S. states would need to be evacuated to avoid the risk of **ash**. Finding transportation, housing, and supplies for all these people would be a massive task, but it would give us the best chance of saving people's lives.

DID YOU KNOW?

Around 800 million people live within 62 miles (100 km) of an **active** volcano. This is close enough to be dangerous in the event of a powerful eruption!

People collect water at an evacuation center in the Philippines in 2018. They were evacuated following high levels of activity from Mount Mayon.

SUPERVOLCANO SAFETY

Scientists are always working on new ideas that could reduce the effects of a volcanic eruption ... or even stop one altogether! At the moment, none of these ideas have been tested on a real eruption, but it's good to have them up our sleeve in case a supervolcano threatens to blow.

One of the most interesting proposals was put together by scientists at the U.S. National Aeronautics and Space Administration (NASA). They suggested that cooling down a volcano, such as Yellowstone, would keep it from erupting. They would do this by digging 6 miles (10 kilometers) down inside the volcano and pumping cold water down inside it. The water would be heated to temperatures of over 660 °F (350 °C) by the sizzling magma inside the volcano. The hot water would then rise back up to the surface, where it could be extracted and maybe even used to generate geothermal power at the same time—a double dose of saving lives and powering lights!

After drilling, hot water would shoot out of the ground like the many geysers in Yellowstone National Park.

This plan may sound good, but there would definitely be risks involved. Digging into a volcano that is close to eruption could speed up the process and get the **lava** flowing! However, scientists believe that this would only happen if they drilled close to the peak of the volcano. Drilling down around the base would be much safer.

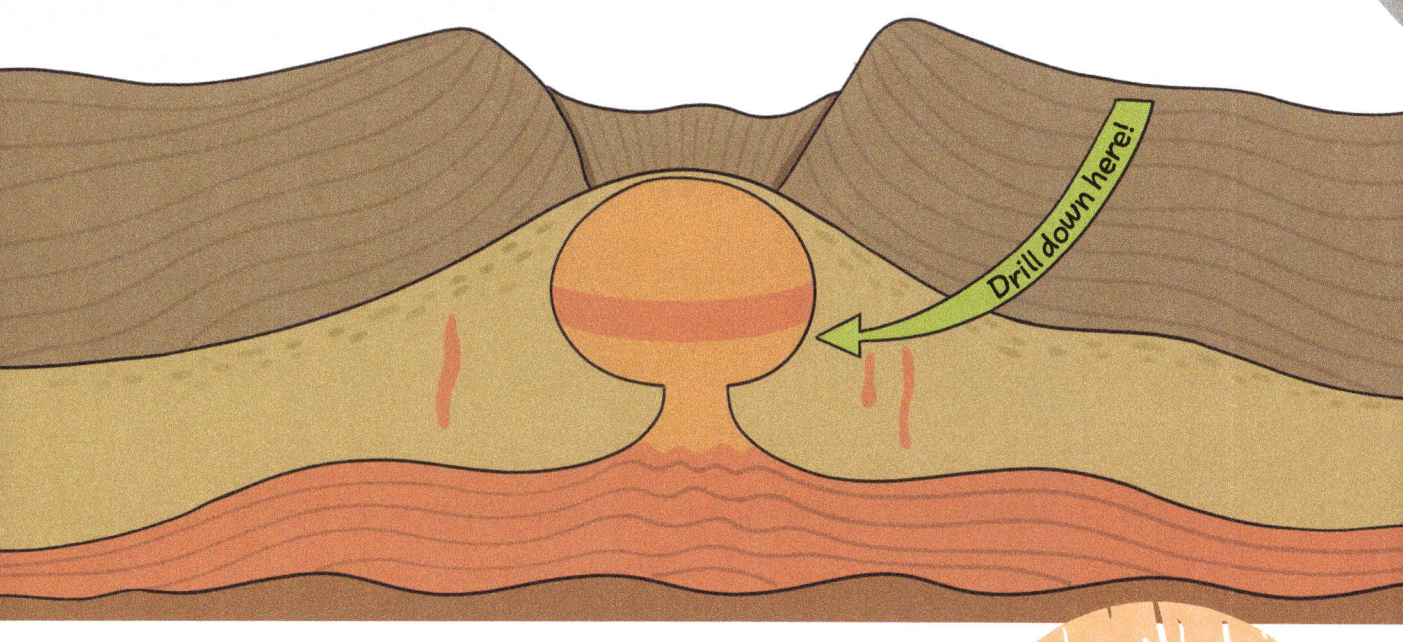

Drill down here!

This plan would also be hugely expensive—$3.5 billion to be precise! Hopefully, some of the money could be made back through generating geothermal energy. And of course, surely that's a fair price to pay for saving lives!

THINK ABOUT IT!

Can you think of another method that could keep a volcano from erupting? It doesn't matter if it sounds silly. Some of the biggest ideas in science today were once thought to be ridiculous!

Conclusion

Although the likelihood of a supervolcano erupting soon is very low, one day Earth will experience another supervolcanic eruption. And in the meantime, other volcanoes are a real threat to many people around the world.

We can't stop the destructive force of a volcano, but we can try to stay out of their way as much as possible! Thanks to the work of **volcanologists**, we know which volcanoes are likely to erupt soon and how they will probably erupt. This means that if a supervolcano were to erupt, most people in the surrounding area would hopefully be **evacuated** in time. However, their homes and businesses would be lost forever under the **lava**, and they'd have to start again from scratch.

FUN FACT!

At least a dozen volcanoes erupt every day! But luckily, these aren't supersized supervolcanoes!

Volcanologists put their lives on the line to research volcanoes.

In the longer term, our whole planet would be greatly affected by the eruption of a supervolcano. We'd probably experience huge changes to our climate and possibly a year or two of **volcanic winter.** This would have a big impact on our ability to grow **crops.** Many people would go hungry, and lives might be lost due to starvation.

However, life on Earth has survived many supervolcanic eruptions. With our scientific knowledge and advanced technology, our chances of making it through are high. In the highly unlikely event that a supervolcano did erupt during our lifetime, we'd probably be OK! For now, we can continue to improve ways of keeping people safe during other eruptions and prepare for the day long in the future when a massive supervolcano does inevitably erupt again.

Little by little, plants and other living things return to the site of an eruption, and life continues.

Summary

So, what exactly would happen if a supervolcano erupted? Check your understanding of the information in this book.

- **Volcanologists** detect high levels of **magma** in a volcano. They predict that a massive supervolcanic eruption will happen soon.

- A supervolcanic eruption takes place.

- The area around the volcano is buried in **lava, pyroclastic flows,** landslides, and **lahars.** Everything is destroyed, including buildings, trees, roads, and other structures.

- A massive cloud of **ash** and gas is released from the volcano.

- If the volcano is near the coast, its eruption might trigger a massive **tsunami.** This wave could create huge amounts of damage along coastlines.

- Ash carried by the wind settles on the ground in areas farther from the volcano. It damages **crops,** blocks **sewers,** and causes breathing problems.

People living within the danger zone of the volcano are **evacuated**.

Airplanes can't fly through the ash cloud. Air travel stops for weeks, months, or even longer.

THINK ABOUT IT!

Do you think it's likely that a supervolcanic eruption will happen in your lifetime? Why or why not?

Eventually, these people may be able to return to the area. However, if their homes and businesses have been totally destroyed, they may need to rebuild their lives elsewhere.

Particles of sulfur in the **atmosphere** reflect sunlight back into space. The temperature on Earth drops, creating a **volcanic winter**. This may last for several years.

Crops don't grow properly without periods of warm weather.

Natural **ecosystems** struggle as wild plants fail to grow in the unnaturally cold weather. Wild animals will go hungry.

Glossary

active—an active volcano might erupt at any time (so watch out!)

ash—black powder left behind after something has burned

atmosphere—the gases around Earth

caldera—a very large crater created when a volcano collapses in on itself after an eruption

citizen science—scientific research done by ordinary people (just like you!) to help scientists with their work

crop—a plant grown for food, such as apples, carrots, or potatoes

crust—the outer layer of Earth

dormant—a dormant volcano isn't currently active but might become active again

ecosystem—all of the living things in an area and the relationship between them

evacuate—to move people from a dangerous place to somewhere safe

extinct—an extinct volcano will never erupt again (whew!)

fertilize—to add something to the soil to make plants grow better in it

geologist—a scientist who studies Earth and the materials from which it is made

geothermal—connected to the heat from inside Earth

ghost town—a town where no one lives anymore (no ghosts though, promise!)

lahar—a huge mudslide created by a volcano that becomes hard when it dries

lava—liquid rock that has reached Earth's surface

magma—liquid rock that is found underground

magma chamber—the underground space inside a volcano where magma builds up before an eruption

mantle—the part of Earth that lies under its crust and around its core

monitor—to watch something carefully over a period of time to see if anything changes

particle—a very small piece of something

pyroclastic flow—a very hot and fast-moving mixture of ash, gas, and solidified lava that comes out of a volcano

sewer—an underground pipe that carries away waste and dirty water from homes and businesses, so that it can be cleaned

tectonic plate—a large piece of Earth's crust

tsunami—a massive wave

volcanic winter—a period of unusually cold weather caused by sulfur particles released by a volcano (see pages 26–27)

volcanologist—a scientist who studies volcanoes

Review and reflect

COMPREHENSION QUESTIONS

Volcanoes 101
- What are volcanoes and how do they form?
- How do volcanic eruptions affect the land around them?

After the eruption
- What is a volcanic winter and what is likely to happen across the planet if we experience one?
- How might the ash from a supervolcano explosion impact travel?

Enormous eruptions
- How do scientists know what they do about supervolcanoes if one hasn't erupted in tens of thousands of years?
- How did the ash from the Mount Toba eruption affect Earth?

Supervolcano safety
- What can we do to protect ourselves from supervolcanoes?
- Who is Jenni Barclay and how does she help protect people from the dangers of volcanoes?

A supersized disaster
- According to British volcanologist Colin Wilson, what is challenging to predict about supervolcanoes?
- What might happen if a supervolcano erupts underwater near a coastline?

Conclusion and summary
- After reading this book and considering what would happen if a supervolcano erupted, what is your biggest takeaway? Why?

MAKE A CHAIN OF EVENTS!

Creating a paper chain can help you explore and visualize how cause and effect relationships can be thought of as a sequence of events.

You'll need:
- Pencil
- Scratch paper
- Pens or markers
- Stapler and staples
- Strips of paper (2 colors, if possible)

Instructions:

1. **Select a focus:** Choose a specific aspect from the book that caught your attention—it could be how a volcano forms and erupts or how the landscape, plants, animals, and people across the world might be impacted by a supervolcano eruption.

2. **Brainstorm causes and effects:** On a sheet of scratch paper, brainstorm and list the causes and effects related to your chosen focus. Think critically about the factors that contributed to or resulted from your focus. You can always look back in the text for ideas!

3. **Write on strips:** Write each cause and each effect on its own strip of paper. If you have different colored paper, use one color for the cause strips and the other for the effect strips.

4. **Create the paper chain:** Organize your strips into causes and effects. Start forming a paper chain to show how a cause leads to an effect. Use the stapler to connect the two strips together. Continue adding cause and effect strips as links in your chain. When you've finished, you should be able to start at the beginning of your chain and read through each chain link in a logical order.

5. **Linking multiple chains:** If your focus has multiple causes or effects, you can create additional chains and link them together to show how complex cause and relationships can be!

Write about it!

Look at the paper chain you created and how the causes link to effects (which in turn link to other causes!). How might breaking a link in the chain impact the overall sequence of events?

World Book, Inc.
180 North LaSalle Street
Suite 900
Chicago, Illinois 60601
USA

For information about other World Book publications, visit our website at www.worldbook.com or call 1-800-WORLDBK (967-5325).

For information about sales to schools and libraries, call 1-800-975-3250 (United States), or 1-800-837-5365 (Canada).

© 2024 (print and e-book) by World Book, Inc. All rights reserved. No part of this publication may be reproduced, stored in a retrieval system, or transmitted in any form or by any means (electronic, mechanical, photocopying, recording, or otherwise) without written permission from World Book, Inc.

WORLD BOOK and the GLOBE DEVICE are registered trademarks or trademarks of World Book, Inc.

Library of Congress Cataloging-in-Publication Data for this volume has been applied for.

What Would Happen If...?
978-0-7166-5448-3 (set, hc.)

A Supervolcano Erupted?
ISBN: 978-0-7166-5450-6 (hc.)

Also available as:

ISBN: 978-0-7166-5456-8 (e-book)
ISBN: 978-0-7166-5462-9 (soft cover)

Staff

Editorial

Vice President
Tom Evans

Editorial Project Coordinator
Kaile Kilner

Curriculum Designer
Caroline Davidson

Proofreader
Nathalie Strassheim

Graphics and Design

Senior Visual Communications Designer
Melanie Bender

Digital Asset Specialist
Rosalia Bledsoe

Written by Izzi Howell
Illustrated by Paula Bossio

Developed with World Book by
White-Thomson Publishing LTD
www.wtpub.co.uk

Acknowledgments

4-5 © N.Minton/Shutterstock; © Biosphoto/Alamy Images
6-7 © Blanscape/Shutterstock; © Dotted Yeti/Shutterstock
8-9 © Marti Bug Catcher/Shutterstock
10-11 © gnoparus/Shutterstock; © Dr Morley Read, Shutterstock; © Erlantz P.R/Shutterstock; © Ahmad Zikri, Shutterstock; © www.sandatlas.org/Shutterstock; © redbrickstock.com/Alamy Images
13 © Dudarev Mikhail, Shutterstock
14-15 © Naeblys/Shutterstock; © ARCTIC IMAGES/Alamy Images; © Arlan Naeg, AFP/Getty Images; © Wirestock, Inc./Alamy Images
16-17 © Science Photo Library/Alamy Images; © GJ-NYC/Shutterstock
18-19 © Ares Jonekson, Shutterstock
20-21 © mauritius images GmbH/Alamy Images; © H. Mark Weidman Photography/Alamy Images
22-23 © Buntoon Rodseng, Shutterstock; © Komkrit Suwanwela, Alamy Images
24-25 © Bignai/Shutterstock; © ZUMA Press, Inc./Alamy Images
26-27 © Elena Berd, Shutterstock; © Mateusz Suska/Shutterstock
28-29 © andreonegin/Shutterstock; © Lukas Gojda, Shutterstock
30-31 © Steeve Raye, Shutterstock; © Oleksiy Mark, Shutterstock; © lri_sha/Shutterstock
32-33 © kan_khampanya/Shutterstock; © Alex Ramsay, Alamy Images
34-35 © tusharkoley/Shutterstock
36-37 © ZUMA Press, Inc./Alamy Images; © Steve Davey Photography/Alamy Images
38 © janaph/Shutterstock
40-41 © ImageBank4u/Shutterstock; © imageBROKER/Alamy Images; © Science History Images/Alamy Images
46-47 © Roi and Roi/Shutterstock

www.ingramcontent.com/pod-product-compliance
Lightning Source LLC
Chambersburg PA
CBHW060943170426
43197CB00023B/2973